ISBN 978-1-5277-3906-2
PIBN 10886015

1 MONTH OF
FREE
READING

at
www.ForgottenBooks.com

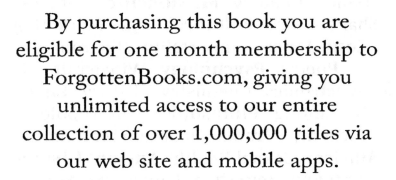

By purchasing this book you are eligible for one month membership to ForgottenBooks.com, giving you unlimited access to our entire collection of over 1,000,000 titles via our web site and mobile apps.

To claim your free month visit:
www.forgottenbooks.com/free886015

UNIVERSITY OF ILLINOIS

190

THIS IS TO CERTIFY THAT THE THESIS PREPARED UNDER MY SUPERVISION BY

ENTITLED

IS APPROVED BY ME AS FULFILLING THIS PART OF THE REQUIREMENTS FOR THE

DEGREE OF

O.B.Wooten.

Instructor in Charge

APPROVED: *Ernst Berg*

HEAD OF DEPARTMENT OF *Electrical Engineering*

197531

Introduction.

Up to the present time very little investigation has been made in regards to ground connections in electrical work, and, therefore, no reliable information is available for this important phase of electrical engineering.

The writers of this work have endeavored to make a study of ground connections which are likely to exist under various conditions, and to determine the best manner of making these connections.

It must be remembered that many variables enter into work of this kind, and hence the conclusions drawn are only general and not specific.

The writers are indebted to Messrs Bell and Coe for valuable data secured in the spring of 1910. Thanks are also due Mr. Wooten for his kind assistance and helpful suggestions.

Description of Grounds.

For this experimental work, the plot of grounds lying directly south of the Electrical Engineering Laboratory was chosen. The earth is partly natural and partly made. Black loam forms the upper two and one-half to three feet, that near the surface showing signs of having been graded in. Below the loam is a stream of yellowish blue clay from six to eight inches in depth. Beneath this, the soil is common blue clay.

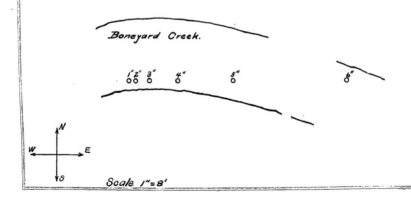

E.E. Laboratory.

Boneyard Creek.

N
W — E
S

Scale 1"= 8'

The grounds were laid out according to diagram and the following arrangement of ground connections chosen.

Number	Kind of terminal	Length in feet.	Kind of point.
1	Pipe	4	Sharpened
2		4	Plain
3		4	
4		4	
5		4	
6		4	
1'		4	
2'		4	
3'		4	
4'		4	
5'		4	
6'		4	Sharpened
1"		4	-
2"		4	Plain
3"		4	
4"		4	
5"		4	
6"		4	
7		1	
8		2	Sharpened
9		6	Plain
10		8	
11		4 salted	
12		6	

- 3 -

Number	Kind of terminal	Length in feet.	Kind of point.
13	Pipe	8 salted	Plain
g		1	
f		2	
e		4	
d		6	
c		8	
b		10	
a		11-1/4	

Plain pipes were those with a circular cross section throughout. The sharpened pipes were plain pipes which had been heated and then one end mashed to a point.

Number 14 is a pipe four feet long, plain joint, with nine number four B and S gage wires, three feet in length, fanned out from the bottom.

Number 15 is a four foot pipe, plain point, with nine number twelve B and S gage wires, three feet in length, fanned out from the bottom.

Number 16 is a copper plate, 2-1/2 by 5 feet by 1/64 inch thickness, buried in coke at a depth of 4 feet.

Number 17 is identical with 16 except without the coke.

The pipes used were the common wrought iron pipes, not galvanized, found on the market in standard sizes.

Work was commenced on the placing of the terminals during the latter part of January 1910 with the frost still in the ground. Considerable difficulty was met in driving

these pipes due to the condition of the ground and to the
joints of the pipes in most cases being plain.

Pipe number 5 was split in driving so was drawn.
The hole was then filled, tamped, and a new pipe driven.
Pipes number 1, 2, 3, 4, 5, 6, 1', 2', 3', 4', 5', and 6'
were all driven the same day, but the ground near the "boneyard" was frozen deeper than that near the laboratory and
may account to a certain extent for the variation in the
resistance of these series of pipes. The main reason, however, for the variation is that pipes 1, 2, 3, 4,⁵/and 6 are
affected by an adjacent water main.

Pipe number 13 struck a rock in being driven so
the contact may be influenced by this.

In locating number 14 and 15, four foot holes were
dug, the pipe and fan put in place and the earth replaced,
tamped, and watered.

Number 16 was dug up during the summer for repairs
and on replacement the earth could not be packed as tightly
as before; consequently there was a mound about eighteen inches
high around the surrounding ground.

The copper plates, number 16 and 17, were fitted
with terminals by soldering number eight B and S gage copper
wire to the centers of the longer sides, near the outer edges,
and running them across to the centre of the plate, being
soldered the entire length. The joints were covered with
two coats of paint, acid and water proof, to prevent electro-
lytic action. They were buried in the manner of number 14

and 15, except seven bushels of coke screenings were packed around number 16, giving a three inch layer of coke all about the plate.

Groups number 14, 15, 16, and 17, at a depth of four feet, lie in the stratum of blue clay mentioned above.

Pipes number 10, 11, and 12 were driven to be used for salt tests, but data was first taken on them as for other pipes.

The pipes numbered from 18a to 22c were driven on April 8, 1910. They are spaced one foot apart, are all four feet in length and range in sizes as follows:-

Numbers			Diameter
18a	18b	18c	3/8"
19a	19b	19c	5/8"
20a	20b	20c	1"
21a	21b	21c	2"
22a	22b	22c	3"

A hole was tapped near the upper edge of each pipe and a pipe plug, bearing about six inches of number twelve copper wire, screwed in. A brass terminal was soldered to the ends of the wires similar to that on the end of a volt-meter lead.

A series of pipes were driven, extending in a north-westerly direction and reaching to Wright Street, with the purpose of testing the variation of resistance with distance. Steam mains, however, so interfered that the readings were worthless for this purpose. The resistance for the first

few feet distance increased, as was to be expected, then the
value decreased gradually, and finally began increasing again.

Method of Operation

The theory connected with this work is an application of Ohm's Law $I = \frac{E}{R}$, where I represents the current,
E the impressed voltage, and R the resistance. The inductance of the earth is negligible. A constant current of
six amperes was always maintained whenever possible.

In the continuous readings simple series connections
were made to each pipe, the return being a water main entering
the east side of the storage battery room. Alternating current was used in order to do away with polarization effect.

Elimination of Errors.

Corrections were made for voltmeter readings by
means of calibration scales for instruments in use, whenever
the error was large enough to warrant such correction. The
ammeter used read correctly.

All curves, except continuous, were plotted from
the average value of several readings. No curve has been
plotted from one single set of readings.

No correction was made for the resistance of the
lead wires since it was negligible. The work is not accur-

ate enough to carry any calculation greater than .1 of an ohm.

Variations of voltage probably caused error in several readings. There is, however, no method of correcting for this.

Outline of Curves

Curve I.

Variation of resistance with depth of pipe.

Curve II.

Variation of resistance with diameter of pipe.

Curve III.

Variation of resistance with distance between two
similar pipes in parallel.

Curve IV.

Variation of resistance with distance, pipes in series.

Curve V.

Variation of resistance with time after salting.

Curve VI and VII.

Tests with direct current.

Curve VIII.

Test with direct current, water poured in pipe.

Curve IX.

Alternating current continuous test, water poured in
pipe.

Curve X.

Variation of resistance with time of year, short pipes.

Curve XI.

Variation of resistance with time of year, long pipes
and plates.

Curve XII.

Curve verifying Curve I.

Curve XIII.

Curve showing variation of resistance with depth,
salted pipes.

Results.

Curve I.

This curve shows that the resistance decreases very
rapidly with increasing depth of pipe up to the length of
about four feet. For pipes longer than six feet, the resistance decreases very slowly, and after a depth of about twelve
feet would probably be nearly constant. This is due, in
part, to the difference in area of contact, but the fact that
the short pipes are buried in the upper loose loam also
accounts for the higher resistance. The curve seems to show
that clay is not a good conductor. If such were the case,
there would be a sudden break downward in the curve instead
of the curve being practically horizontal.

Curve II.

Here the effect of size of pipe is shown, the larger
the pipe the less the resistance, due to the greater area of
contact. The decrease of resistance, however, does not con-
tinue indefinitely. It decreases quite rapidly up to a
diameter of two inches, after which it decreases very slowly

and would probably have become constant with larger diameters.
Curve III.

These curves for parallel circuits show that the
resistance decreases rapidly with increase of distance between
pipes, up to about eight feet, after which the decrease is
slow and gradual, and would probably become constant eventually.
This is due to the greater drop in the earth when the circuits
are close together.   The current density varies inversely
with the distance between terminals, caused by the interference
of flow lines between pipes.
Curve IV.

For series circuits the ground resistance increases
rapidly up to about ten feet after which it becomes nearly
constant.   This result is verified by the data taken for two
separate sets of pipes in different parts of the plot of
ground.   An interesting part here is that those pipes near
the "Boneyard" have a resistance of about twice those near the
laboratory.   This is probably due to location of the water
main.
Curve V.

The effect of using artificial means to increase the
conductivity of the earth and contact, is well shown by these
curves.   The resistance of the pipes after being salted immed-
iately falls several ohms.   The decrease of resistance is
rapid for the first several minutes, more gradual for the next
couple of hours, and about constant from then on.

On April 27, 1911, pipes number 6' and 11 were dug
up and examined.   Pipe 11 was salted, but 6' was not.   The

Salted pipe was slightly corroded, but the other had not been effected.

Curve VI and VII.

These curves show the effect of continuous use of direct current. The increase of resistance is very gradual for the first hour, after which it increases very rapidly for a considerably longer time.

Curve VIII.

This curve shows the same results as VI and VII, and in addition shows the effect of pouring in water. The water moistens the earth thoroughly, increases the conductivity, and hence decreases the resistance.

Curve IX.

This curve shows the effect of using an alternating current continuously, then increasing the conductivity by pouring water in the pipe. It will be noticed that the resistance decreases slowly for about an hour, then increases very rapidly. This condition is probably brought about by the chemical action of the salts in the soil.

A three foot pipe with a number of holes drilled in sides was used for this test. The pipe was two inches in diameter and had sharpened ends to prevent dirt getting inside the pipe.

Curves X and XI.

Curve X shows that the resistance of short pipes is not at all constant but varies greatly with the season of the year.

Curve XI shows that all the pipes and plates are

affected by the time of the year, but not to such an extent as the one and two foot pipes in Curve X. The resistance increases with decrease of temperature, and decreases with rain fall.

The resistances of the pipes in all cases are much higher than those of the plates.

Terminals 16 and 17 are identical copper plates buried in the earth, number 16 being surrounded with a three inch layer of coke screenings. The resistance of both is quite uniform, that of number 16 being several ohms lower.

Pipe number 14 with the wires fanned out from the bottom and the salted pipe number 11 both remain fairly constant, the former, however, having the least resistance. Curve XII.

This curve is simply a verification of Curve I. Curve XII.

This is a curve showing the variation of resistance with the depth of pipe, the pipes being salted. Not enough data is available to enable any definite conclusions being drawn, only three depths being used. It would seem from the curve that the resistance decreases very slowly with the depth.

Conclusions.

The results show that the resistance decreases
with:-

1. Length of pipe up to about twelve feet.
2. Diameter of pipe up to about three inches.
3. Distance between pipes in parallel up to about
    twelve feet.
4. The addition of artificial matter of high con-
    ductivity.
5. That the resistance increases with the distance
    between pipes in series up to about ten feet.

For a good cheap, efficient ground with resistance
nearly constant, one such as number eleven would be very satis-
factory. This is a plain, two inch, four foot pipe, salted.
The resistance could be reduced still farther by using two
such pipes in parallel at a distance of about eight feet.

Printed by BoD™in Norderstedt, Germany

9 781527 739062